101 Fascinating Facts
About Your World

Larry Esquivel

Printed in the United States of America

To the best social studies teachers in the world:

Mr. Matt Ehresman

Mrs. Sheri Haag

Mr. Todd Sheely

Our world is truly an amazing and fantastic place. I am always surprised to see how many stunning things there are to experience and enjoy. Spectacular places built by man, like the Great Wall of China and the Pyramids of Egypt, and the beautiful creations of nature, like Mt. Everest and Niagara Falls.

I enjoy reading fun and interesting facts about the world. If I do not feel like reading a long, boring novel or textbook, I visit my bookshelf for books of facts and World Records. I have a marvelous time asking trivia questions or telling my friends the interesting facts I have learned.

It is important to learn about Earth. It is the place we live in, our home. One does not have to take a geography class to learn about our planet. Find and read a fact book about it and then go and discover it yourself. I hope you enjoy these interesting facts and I am pretty sure you will learn one or two things about "your world".

-Larry Esquivel

1. Water covers approximately 71% of the earth's surface, and land the remaining 29%.

2. A year on Earth it is actually 365.2564 days. It's this extra .2564 days that creates the need for leap years. That's why we tack on an extra day in February every year divisible by 4, such as 2004, and 2008.

3. The Earth is not actually round in shape; in fact it is geoid. This simply means that the rounded shape has a slight bulge towards the equator.

4. Earth is the only planet whose name has not been derived from the Roman or Greek mythology. The name has originated from the 8th century Anglo Saxon word "*Erda,*" which means "ground or soil."

5. Earth moves around 18.5 miles in a second in the orbit around the sun.

6. The distance from the earth to the sun is 93 million miles.

7. Gravity is what keeps earthlings alive, and from flying off into space as the earth spins around.

8. Our Planet is the third closest to the Sun in our Solar System, the average distance being 150 million Km.

9. The Atacama Desert in Chile did not have any rainfall for 400 years until 1971. It is considered the driest plateau across the world.

10. The oldest known rocks, found in Western Australia, are about 3,200 million years old.

11. The highest waterfall on Earth is the Angel Falls, they are 3,212 feet high.

12. The Earth has a surface area of 510,101,000 sq. km's.

13. At 840,000 square miles (2,175,590 km^2), Greenland is the world's largest island.

14. The Amazon rainforest produces more than 20% the world's oxygen supply.

15. Canada has more lakes than the rest of the world combined.

16. Canada is an Indian word meaning "Big Village."

17. The first city to reach a population of 1 million people was Rome, Italy in 133 B.C.

18. Siberia contains more than 25% of the world's forests.

19. Spain literally means 'the land of rabbits.'

20. The deepest hole ever made in the world is in Texas. It is as deep as 20 empire state buildings but only 3 inches wide.

21. The Earth is estimated to be more than 4.5 billion years old.

22. Earth has one moon. It is about one-quarter the size of the Earth.

23. The moon orbits the Earth, taking about 27days and 7 and half hours to move around the Earth once.

24. The Earth is made up of several layers. There is an outer solid crust, then a mantle, then a liquid outer core, and finally a solid inner core.

25. There are ten types of clouds and they are found at different levels of the atmosphere. The three main kinds of clouds are cirrus, stratus, and cumulus.

26. A glacier is a river of frozen fresh water, and made of layers of snow that turn to ice.

27. Only 11 percent of the earth's surface is used to grow food.

28. Breeze carries about 100 million tons of sand particles around the earth yearly. That means if you live in America-you could have sand that came from the Gobi desert in China.

29. Earth is estimated to weigh 6,585,000,000,000,000,000,000,000 tons.

30. Earth is tipped at 23 and 1/2 degrees in orbit. That axis is what causes our seasons.

31. Earth is tipped at 23 and 1/2 degrees in orbit. That axis is what causes our seasons.

32. The Earth's atmosphere is composed mainly of nitrogen (77%), oxygen (21%), argon (.93%), and carbon dioxide (0.03%).

33. Earth is the only presently known planet in the Solar System to support life. The earliest fossil evidence for life dates back 3.5 billion years ago.

34. No one really knows how our planet came into being, but one common theory is that the earth developed form a swirling mass of rock and gas. 100 million years ago, the single landmass (called Pangaea) broke into the continents we know today.

35. It takes light 8 minutes and 18 seconds to travel from the sun to the Earth.

36. Earth is the 5th largest planet almost 2 million square miles.

37. There are 10 Million Species of life on Earth.

38. Antarctica is the coldest recorded place on earth.

39. Deepest place in the Ocean is Marianas Trench in Marianas Islands; the bottom is 35,813 ft or about 7 miles and is called "Challenger Deep."

40. There are five oceans on Earth. They are: the Pacific Ocean, Atlantic Ocean, Arctic Ocean, and the Indian Ocean.

41. The Earth has a population of 6 billion human beings.

42. There are 24 time zones on Earth.

43. There are more than 6,800 different languages, and 10,000 religions with numerous denominations.

44. The highest tides in the world are at the Bay of Fundy, which separates New Brunswick from Nova Scotia.

45. Rainforests once covered 14% of the earth's land surface, now they only cover 2% and it could all be consumed in 40 years.

46. 121 prescription drugs sold worldwide come from plant-derived sources in the Rainforest.

47. Tropical rainforests are the world's oldest ecosystems.

48. There are 2 types of Rainforest, the temperate and the tropical.

49. 25% of the world's species could be lost by the end of this decade.

50. Every ton of paper that is recycled saves 17 trees.

51. Each person throws away approximately four pounds of garbage every day.

52. One-third of all energy is used by people at home.

53. 14 billion pounds of trash is dumped into the ocean every year.

54. One gallon of motor oil can contaminate up to 2 million gallons of water.

55. The Arctic ice pack has lost about 40% of its thickness over the past four decades.

56. If Antarctica were to melt, the sea level would rise over 200 feet.57. Asia has a population of 3.3 billion people, nearly three fifths of the world's total population.

57. Mt. Everest is the highest point on Earth; it is 29, 029 feet above sea level, and the lowest point is the Dead Sea; it is 1385 feet below sea level.

58. The highest temperature ever recorded on Earth was 136 degrees Fahrenheit in 1922, in El Azizia, Libya. The coldest was -129 Fahrenheit in Vostok, Antarctica.

59. The Kuroshio Current, off the shores of Japan, is the largest current in the world.

60. Antarctica makes 90% of the total ice on Earth.

61. 90% of all volcanic activity occurs in the oceans.

62. 3,000 fruits are found in the rainforests; currently we use 200, rainforest Indians use 2,000.

63. The United States National Cancer Institute has identified 3000 plants that are active against cancer cells. 70% of these plants are found in the rainforest.

64. Each person uses about 12,000 gallons of water every year.

65. Each gallon of fuel releases 20 pounds of carbon dioxide into the air.

66. It takes 90% less energy to recycle aluminum cans than to make new ones.

67. The Andes are the longest mountain range in the world; it extends more than 4,000 miles through seven countries down the western portion of South America.

68. American swimming pools contain enough water to cover the city of San Francisco with a layer of water about seven feet deep.

69. In one year, we generate enough hazardous waste to fill the New Orleans Superdome 1,500 times over.

70. From a distance, Earth would be the brightest of the 9 planets. This is because sunlight is reflected by the planet's water.

71. The center of the Earth, its core, is molten. This means that it is liquid rock which sometimes erupts onto the surface through volcanic eruptions. This core is 7,500°c hotter than the surface of the Sun.

72. The earth rotates on its axis more slowly in March than in September.

73. To escape the Earth's gravity and get out into space, a rocket has to travel at a speed of 25,100 mph or 11.18 km/sec. That's about 386 times faster than your parents are allowed to drive on a U.S. highway!

74. Lake Bosumtwi in Ghana formed in a hollow made by a meteorite.

75. Off the coast of Florida there is an underwater hotel. Guests have to dive to the entrance.

76. The oldest living tree is a California Bristlecone pine named "Methuselah." It is about 4,600 years old.

77. The Ancient Egyptians worshipped a sky goddess called Nut.

78. The United States uses 29% of the world's petrol and 33% of the world's electricity.

79. Tibet is the highest country in the world. Its average height above sea level is 4500 meters.

80. Russia is only two miles from Alaska.

81. The Sahara Desert is about the same size as the United States.

82. A clock runs faster on a tall mountain than at sea level.

83. Every day, more than 16,000 children die from hunger-related causes. This is one child every five seconds.

84. Currently, one in five people in the world survive on less water per day than is used to flush a toilet.

85. The United Nations flag is the only flag that may fly above an USA flag on an American flagpole.

86. Libya's flag is the only flag which is all one color (green) with no writing or decoration on it.

87. In 1783 and eruption in Iceland created so much dust it blocked out the sun over Europe for a short time.

88. There is a city called Rome in every continent.

89. The number of births in India each year is greater than the entire population of Australia.

90. There are more plastic lawn flamingos in the US than real ones.

91. The letter "Q" is the only letter in the alphabet that does not appear in the name of any of the United States.

92. The Arctic Ocean is the world's smallest ocean.

93. Earth's oceans average two miles deep.

94. The Earth's closest planetary neighbors are Venus and Mars.

95. Earth is slowing down. Every few years, an extra second is added to make up for lost time. Millions of years ago, a day on Earth will have been 20 hours long. In a few million years, a day will be 27 hours on Earth.

96. The average distance between Earth and Moon is 238,857 miles (384,403.1 km).

97. It is illegal to catch fish with your bare hands in the state of Kansas.

98. There are 195 countries in the world.

99. The most dangerous animal in the world is the common housefly. Because of their habits of visiting animal waste, they transmit more diseases than any other animal.

100. Luxembourg is the wealthiest nation in the world. The United States is the 4th.

101. The temperature of the Earth's interior increases by 1 degree every 60 feet down.

About the Author:

Larry Esquivel lives with his parents in Greenwood, Indiana. He has published a variety of articles in many local magazines and newspapers. Larry enjoys reading, playing piano, and writing. He is also a volunteer at a public library in Greenwood where he spends most of his free time serving the community. Larry also likes going to bookstores and spend some time alone to find some books to read.

www.ingramcontent.com/pod-product-compliance
Lightning Source LLC
Chambersburg PA
CBHW060342290526
45791CB00004B/1503